Shadows

By Aimee Louise

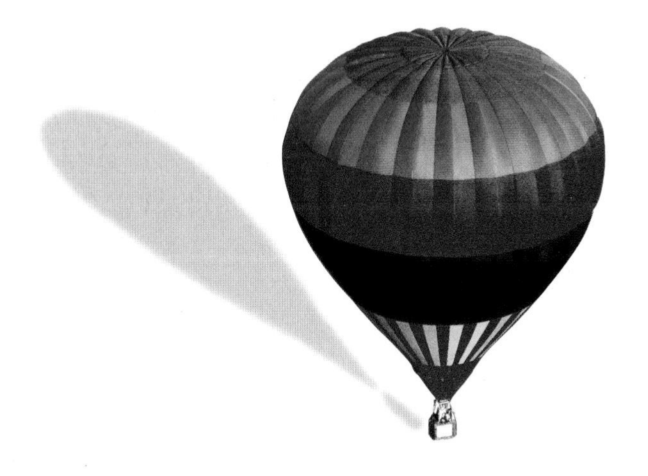

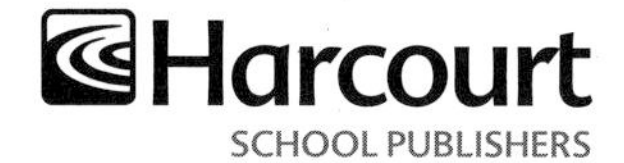

SCHOOL PUBLISHERS

Orlando Austin New York San Diego Toronto London

Visit *The Learning Site!*
www.harcourtschool.com

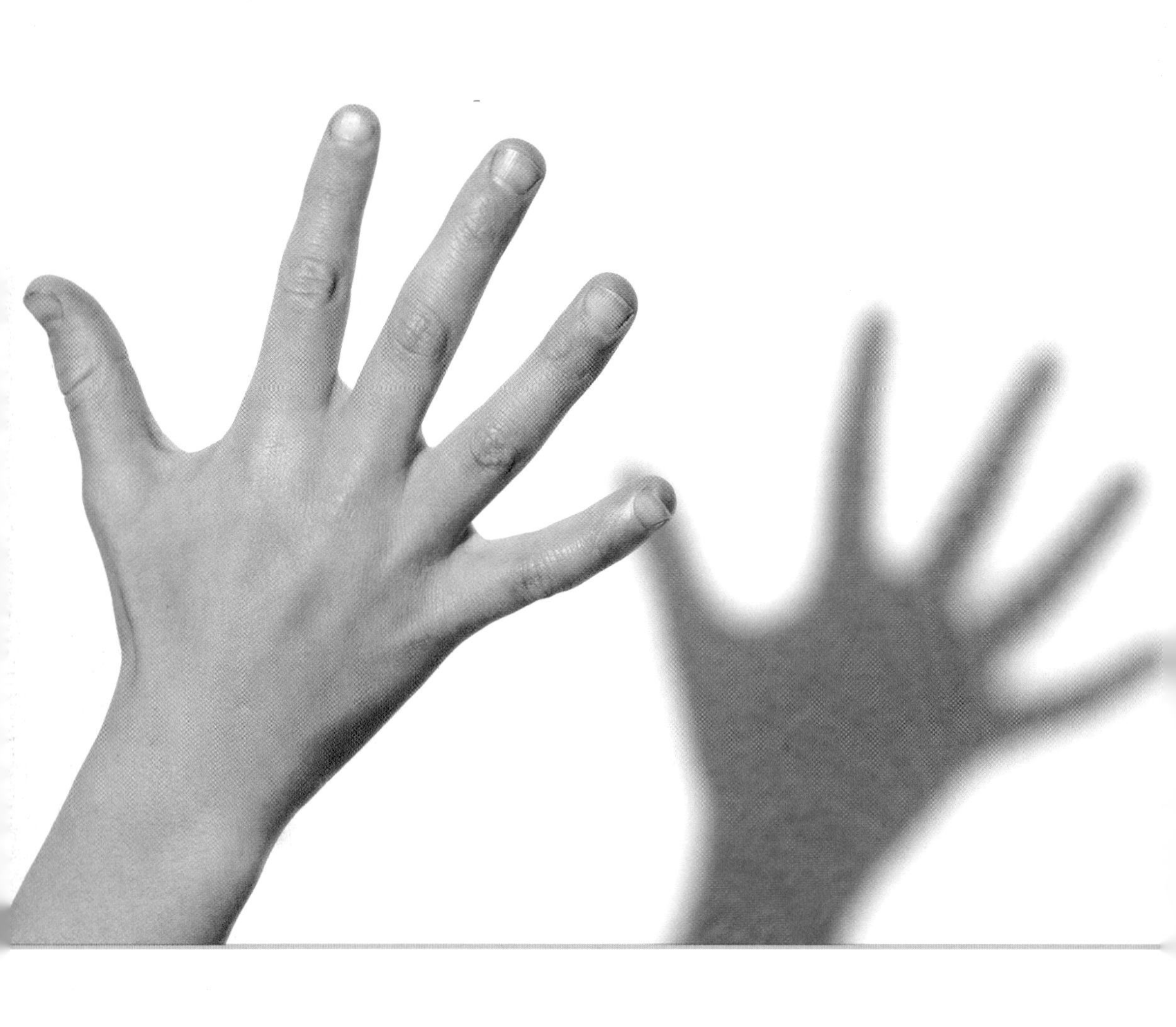

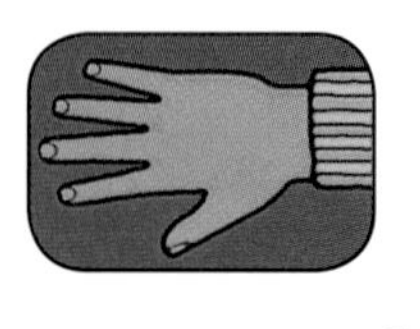

a hand shadow

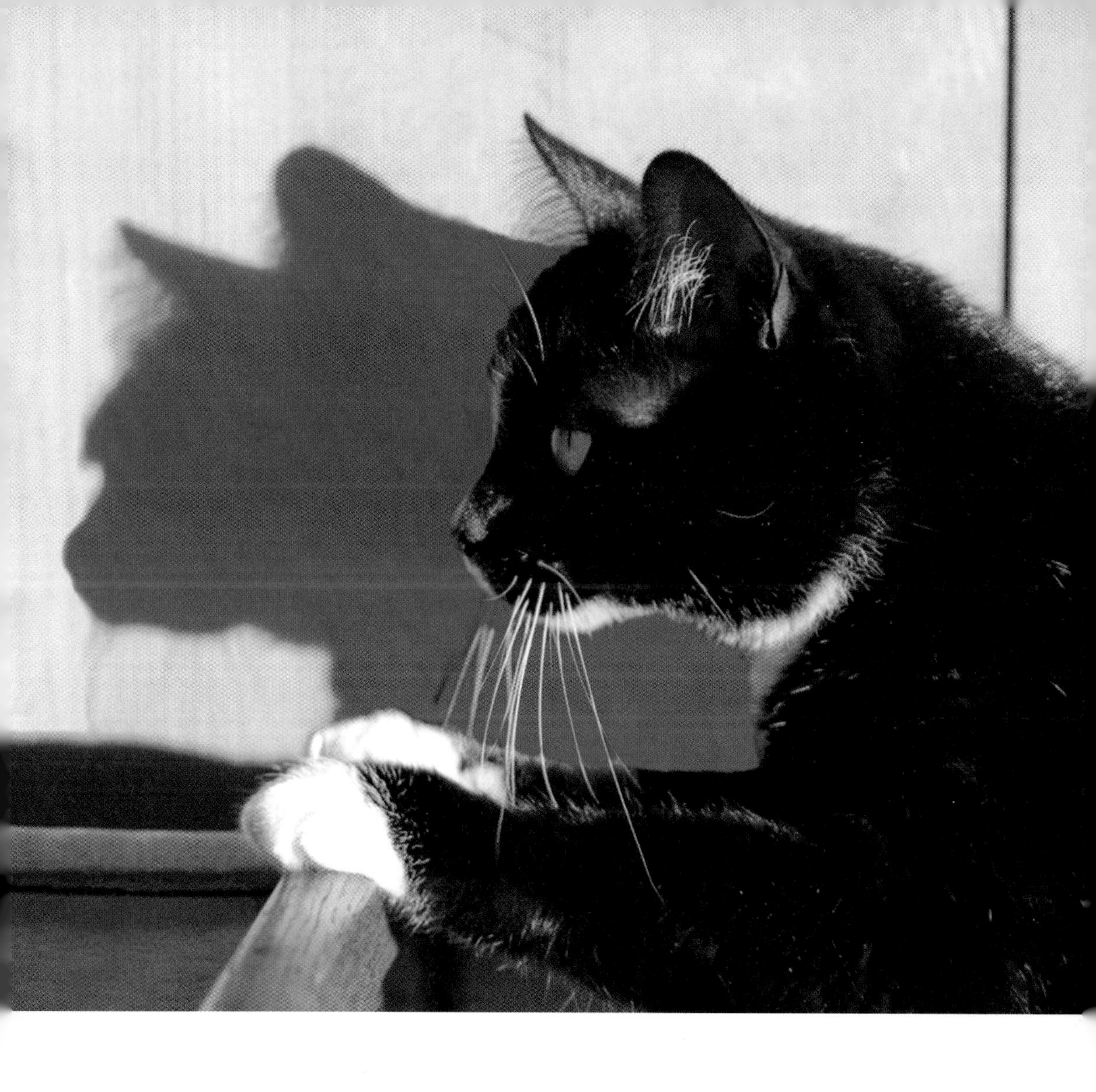

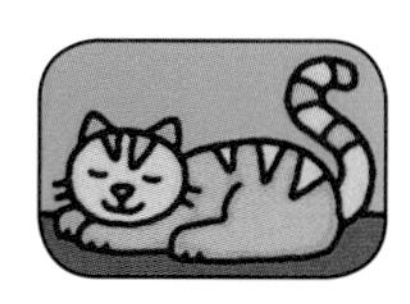

a cat shadow

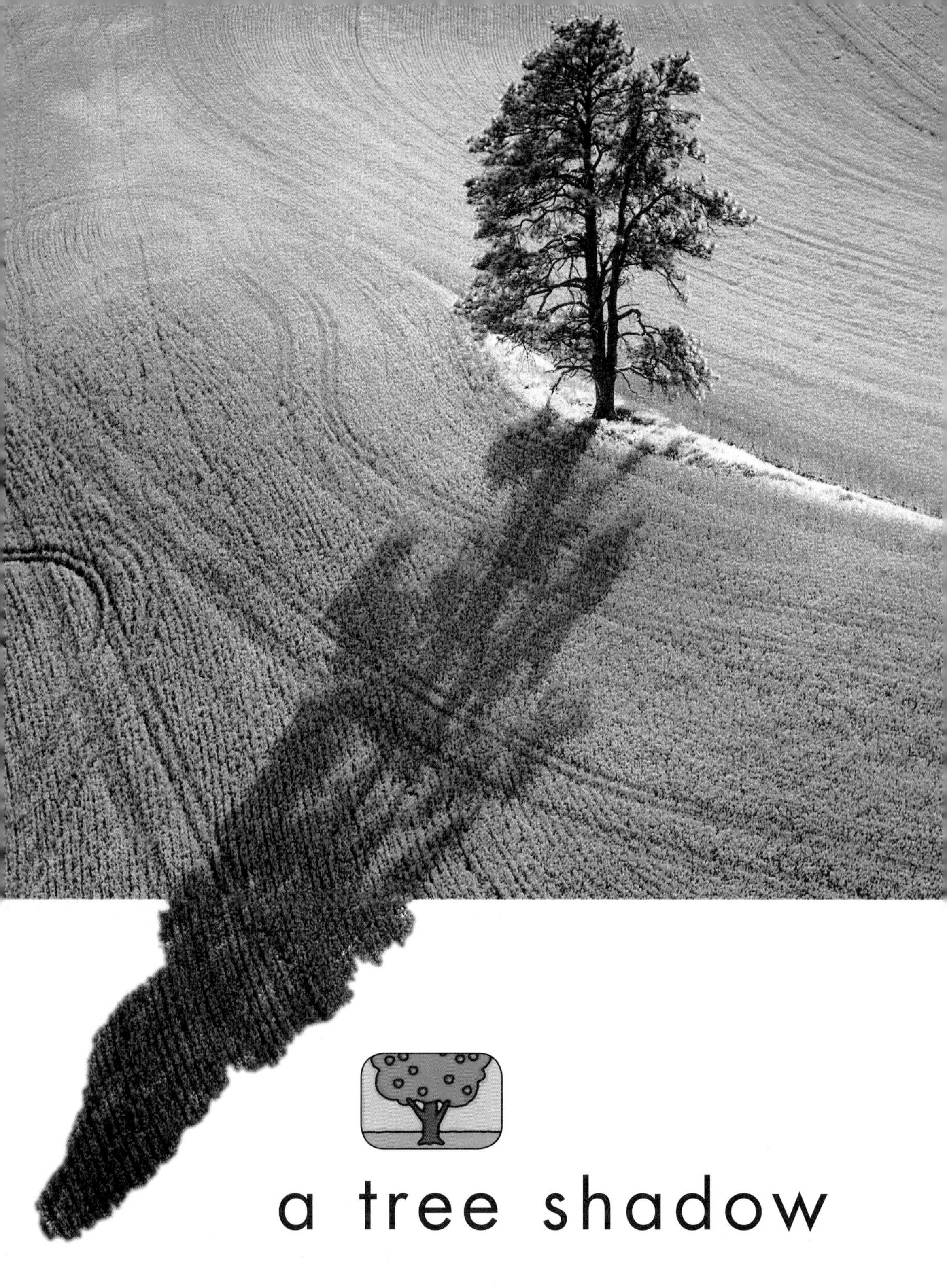

a tree shadow

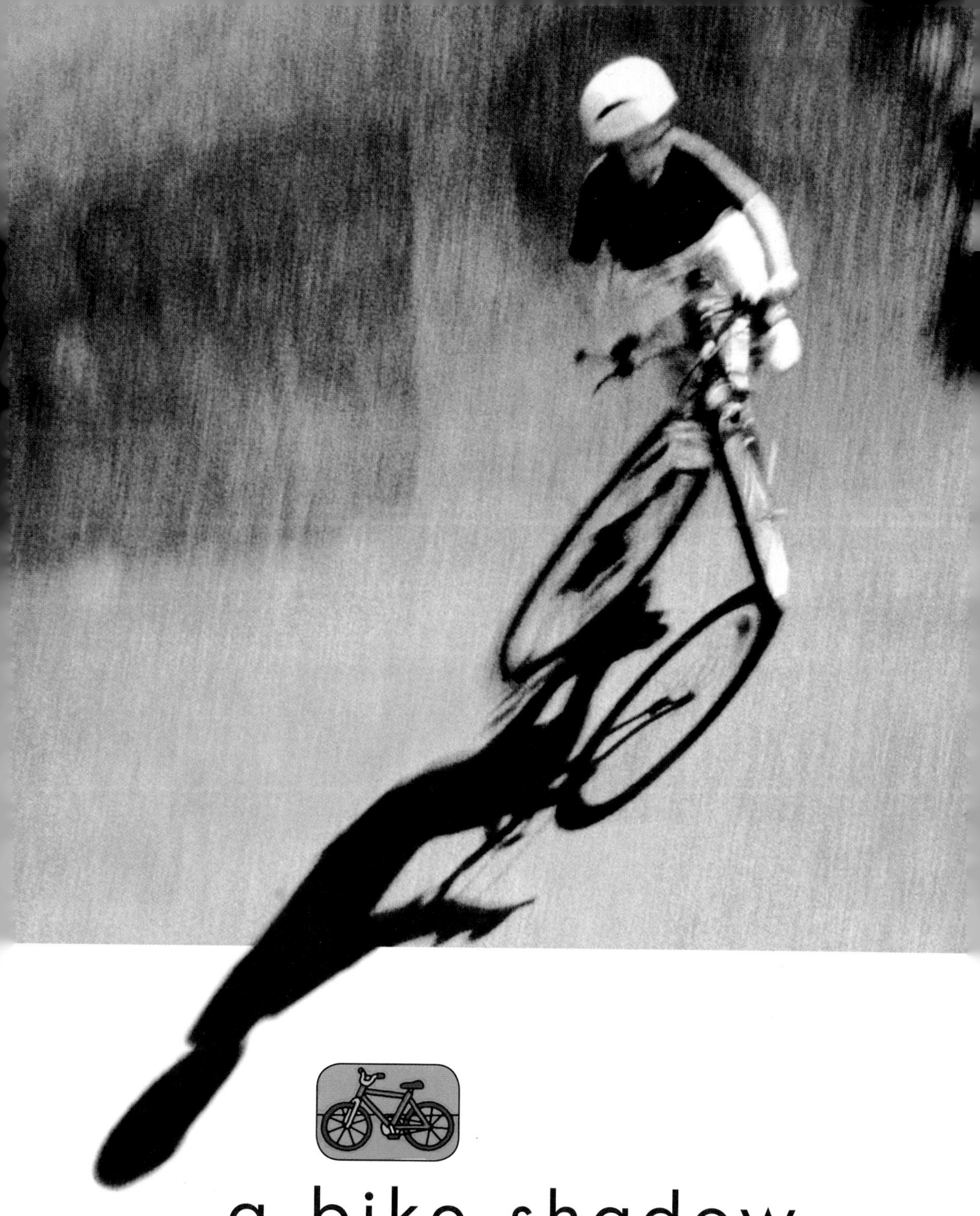

a bike shadow

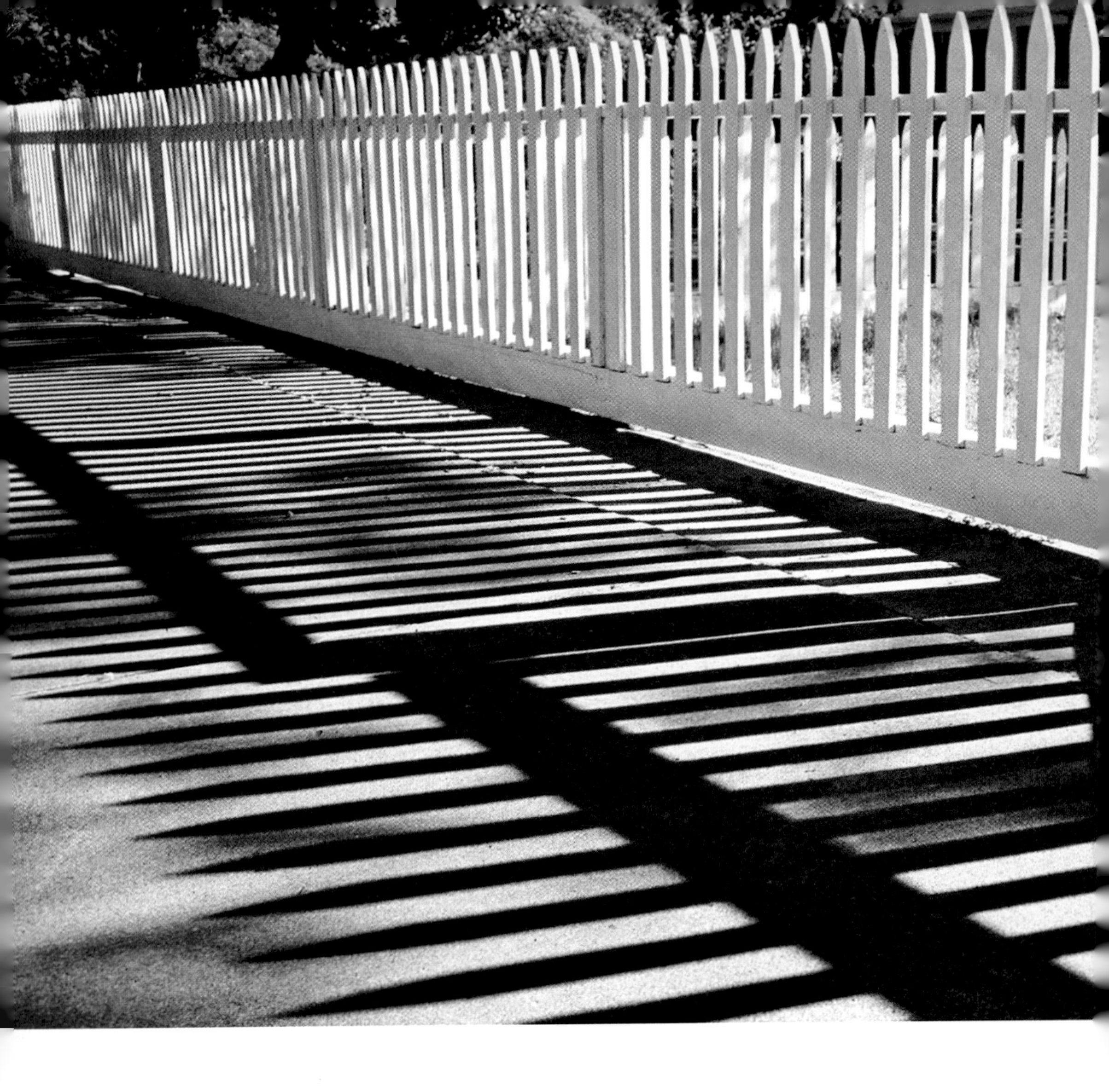

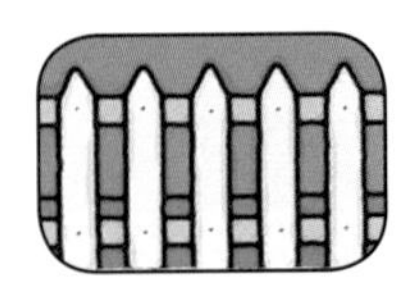

a fence shadow

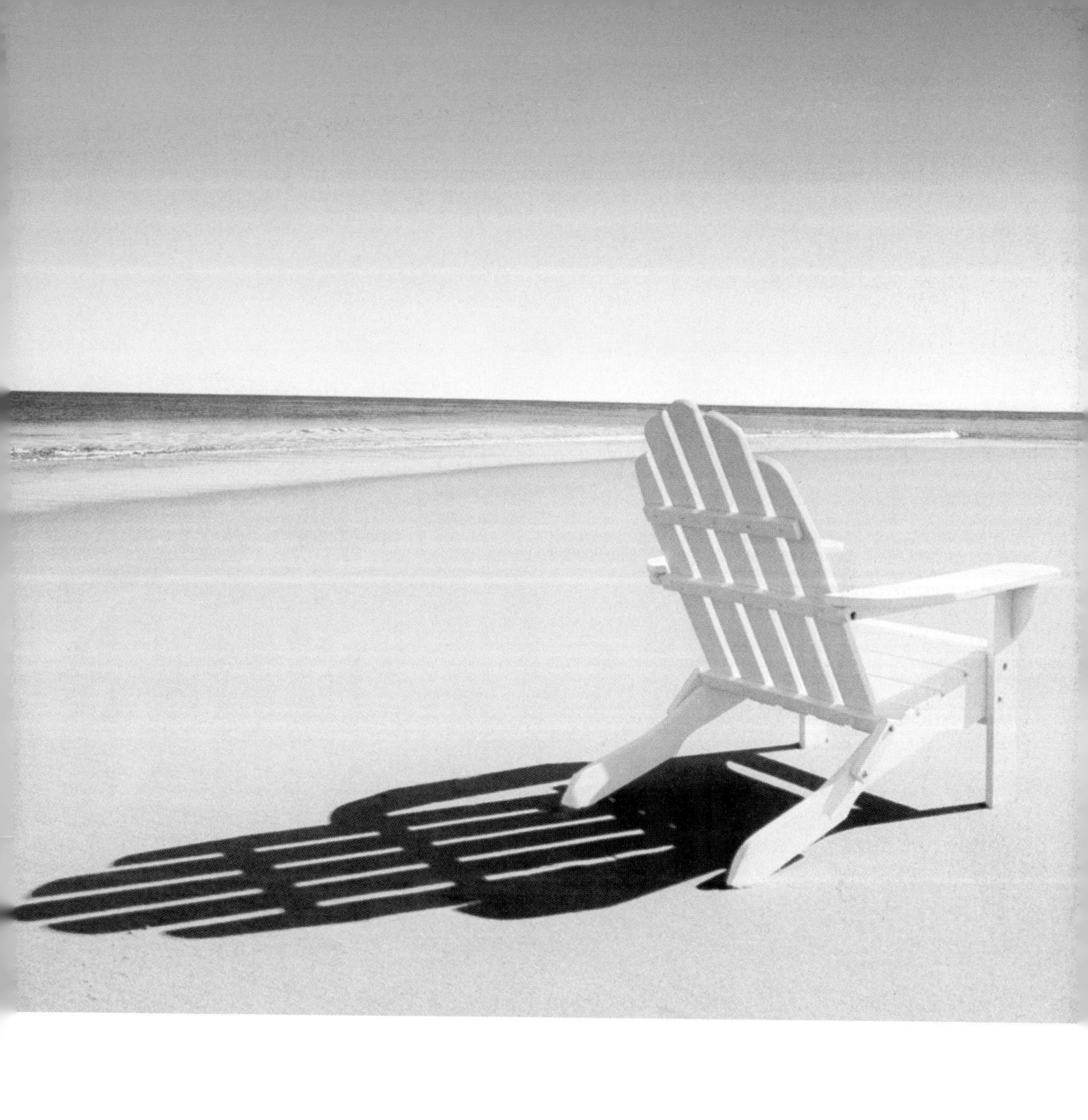

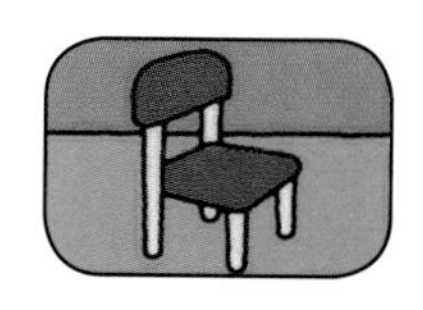

a chair shadow

My shadow!

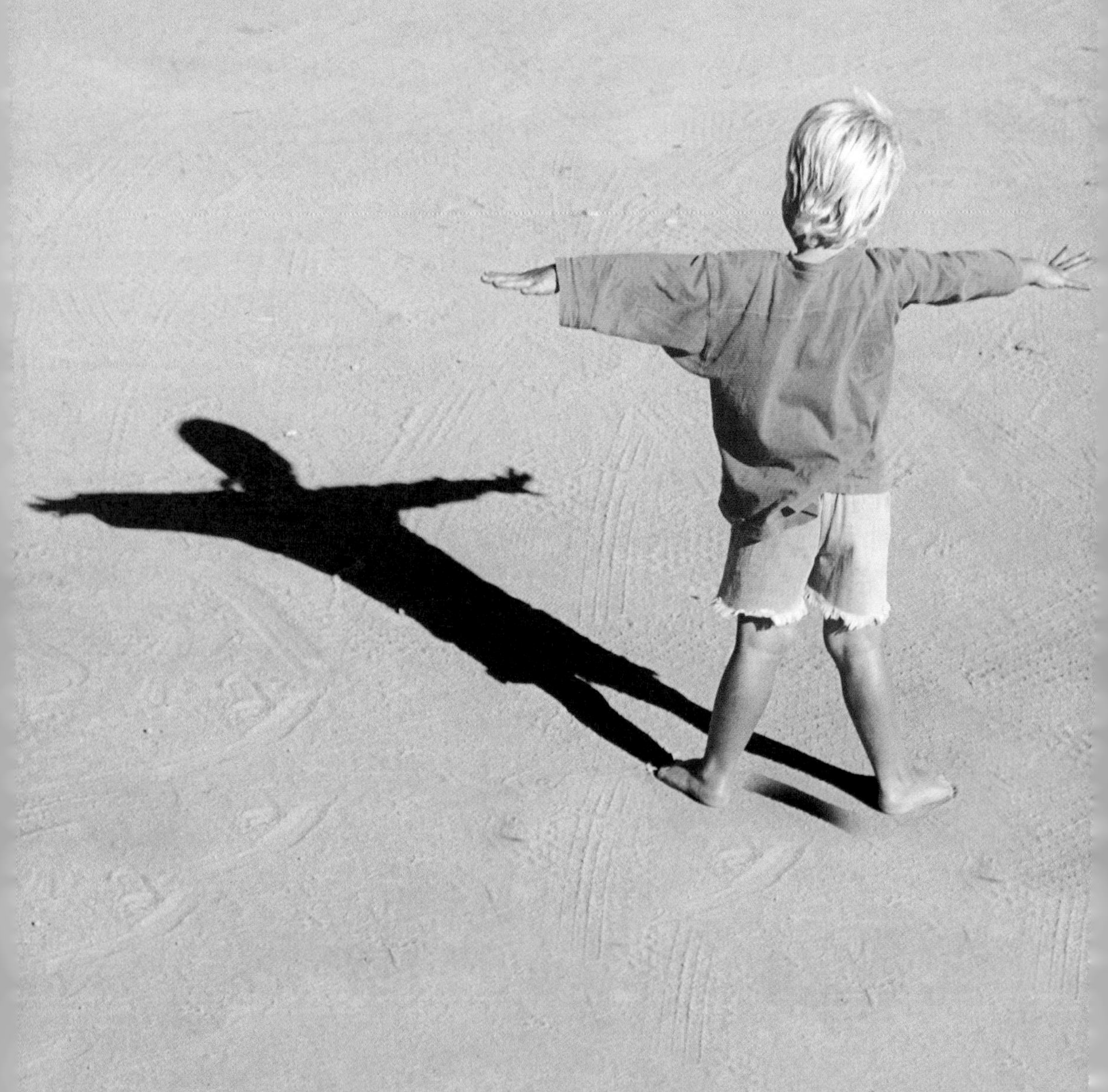